Polar Babies

Mary Elizabeth Salzmann

Consulting Editor, Diane Craig, M.A./Reading Specialist

Sandcastle

An Imprint of Abdo Publishing
abdobooks.com

abdobooks.com

Published by Abdo Publishing, a division of ABDO, PO Box 398166, Minneapolis, Minnesota 55439. Copyright © 2020 by Abdo Consulting Group, Inc. International copyrights reserved in all countries. No part of this book may be reproduced in any form without written permission from the publisher. SandCastle™ is a trademark and logo of Abdo Publishing.

Printed in the United States of America, North Mankato, Minnesota

052019
092019

THIS BOOK CONTAINS
RECYCLED MATERIALS

Design: Christa Schneider, Mighty Media, Inc.
Production: Mighty Media, Inc.
Cover Photograph: Shutterstock Images
Andrey Zvoznikov/Ardea.com, pp. 4 (lemming), 18–19; Hiroya Minakuchi/National Geographic Image Collection, pp. 4 (orca), 14–15; Paul Nicklen/National Geographic Image Collection, pp. 5 (walrus), 20–21; Shutterstock Images, 4 (penguin), 5 (polar bear, puffin, reindeer, seal), 6–7, 8–9, 10–11, 12–13, 16–17, 22 (flowers, grass, sky, snow)

Library of Congress Control Number: 2018966951

Publisher's Cataloging-in-Publication Data
Names: Salzmann, Mary Elizabeth, author.
Title: Polar Babies / by Mary Elizabeth Salzmann
Description: Minneapolis, Minnesota : Abdo Publishing, 2020 | Series: Animal babies
Identifiers: ISBN 9781532119613 (lib. bdg.) | ISBN 9781532174377 (ebook)
Subjects: LCSH: Animals--Polar regions--Juvenile literature. | Animal babies--Juvenile literature. | Polar animals--Juvenile literature.
Classification: DDC 599.7--dc23

SandCastle™ Level: Emerging

SandCastle™ books are created by a team of professional educators, reading specialists, and content developers around five essential components—phonemic awareness, phonics, vocabulary, text comprehension, and fluency—to assist young readers as they develop reading skills and strategies and increase their general knowledge. All books are written, reviewed, and leveled for guided reading and early reading intervention programs for use in shared, guided, and independent reading and writing activities to support a balanced approach to literacy instruction. The SandCastle™ series has four levels that correspond to early literacy development. The levels are provided to help teachers and parents select appropriate books for young readers.

EMERGING • BEGINNING • TRANSITIONAL • FLUENT

Contents

Polar Babies

Can you find these baby polar animals in this book?

baby lemming

baby orca

baby penguin

baby polar bear

baby puffin

baby reindeer

baby seal

baby walrus

See the baby polar bear.

See the
baby
reindeer.

See the baby puffin.

See the baby penguin.

See the

baby

orca.

See the
baby
seal.

See the

baby

lemming.

See the
baby
walrus.

What Else Did You See?

flowers

grass

sky

snow

Index

Teacher's Guide

ATOS: 0.9   GRL: A   Word Count: 33

High-Frequency Words

See   **the**

Content Words

baby, lemming, orca, penguin, polar bear, puffin, reindeer, seal, walrus

Before Reading

- Tell students that the title of the book is *Polar Babies*.
- Summarize the content of the book. Explain what "polar" means.
- Have students look through the book. Ask them what they see in the pictures.
- Choose a few new vocabulary words. Have students predict what letter each word starts with. Then have them find the words in the book.

After Reading

Ask students questions about the book's content, such as:

- What animals did you see in the book?
- How could you tell that the animals live where it is cold?
- Have you ever been to very cold place? What was it like?
- What other animals would you like to read about?